The Big Bang and the Fine-tuned Universe

Robert A. Alston

Seattle   Discovery Institute Press   2024

Description

In easy-to-understand language, former NASA special projects engineer Robert Alston tackles cosmology's profoundest questions: Where did the universe come from, and how were its laws and constants finely tuned to allow for life? From Albert Einstein's biggest blunder to the perfect parameters that allow fragile life to persevere, this mini-book explores how astronomy and physics point to a cosmic architect.

Library Cataloging Data

The Big Bang and the Fine-tuned Universe by Robert A. Alston

34 pages, 6 x 9 in.

ISBN-13 Paperback: 978-1-63712-049-1, Kindle: 978-1-63712-051-4, EPub: 978-1-63712-050-7

BISAC: SCI015000 SCIENCE / Space Science / Cosmology

BISAC: SCI034000 SCIENCE / History

BISAC: SCI075000 SCIENCE / Philosophy & Social Aspects

Publisher Information

Discovery Institute Press, 208 Columbia Street, Seattle, WA 98104

Internet: https://discovery.press/

Published in the United States of America on acid-free paper.

First Edition, First Printing, August 2024.

CONTENTS

THE BIG BANG AND THE
FINE-TUNED UNIVERSE

Robert A. Alston

EVER WONDER HOW PHYSICISTS EXPLAIN WHERE THE UNIVERSE came from? For a long time the go-to explanation was that the universe didn't come from anywhere because it simply always existed. But then a funny thing happened. Scientific discoveries revealed that our universe was not always here. It came into existence about 13.8 billion years ago. Sit back while I tell you the story of how we came to this discovery.

Since the invention of the telescope, humans have peered deep into the night sky and seen the wonders of the cosmos previously hidden from us. The more the telescope advanced, the farther we looked and the more we saw. The farther and farther we looked the more it all looked the same: stars and gas clouds. It seemed like it would never end. This reaffirmed the view of many scientists that the universe was infinitely large and infinitely old. At the turn of the twentieth century, this was the conventional wisdom in astronomy.

Then, in 1915, German physicist Albert Einstein changed everything with his general theory of relativity. After working out his calculations (with pencil and paper, I might add), he noticed something odd. The calculations suggested that the universe was either expanding or contracting. This was strange. Einstein thought it had to be a mistake. Like his colleagues at the time, he was certain that the universe was static and eternal. But if it was expanding, it was hardly static, and it was hard to see how an expanding universe could have been expanding forever. So,

Figure 1. Albert Einstein.

he did what any good physicist would do. He fiddled with his equation! Einstein scratched into it another number—a "constant"—which would fix his problem, salvaging the static, eternal universe model.

This seemed to solve the problem, but he would later call this change to his equation the "biggest blunder" of his life.

Hubble Trouble

In 1923, at Mt. Wilson Observatory in California, Edwin Hubble was performing the mundane routine of observing nebulae, objects that appeared extended and fuzzy through a telescope, with the largest telescope in the world at that time, and he discovered that some of these were in reality other galaxies. At the same time, this discovery showed that our Milky Way is also a galaxy—not the whole of the observable universe but simply one galaxy among many in the universe.

As if that weren't amazing enough, he noticed a few years later something peculiar about the light emitted from these galaxies. The farther away a galaxy was from us, the more the color of its light was shifted to red. This told him that these galaxies were moving away from us, and that the farther away they were, the faster they were moving away. It appeared as though the universe was expanding, and indeed it was.

When Einstein caught wind of this, he traveled to California to see this phenomenon for himself. What he saw forced him to erase the mathematical addition to his equation and concede that in fact the universe was expanding.

But it wasn't just strange to his way of thinking. The idea of an expanding universe had profound implications. Imagine we are watching the universe on television and can fast-forward and rewind through time like we can with movies at home. What would it look like if we were to rewind and keep rewinding at high speed? If the universe is expanding over time, as Hubble realized, then if we rewound our movie of the universe backwards in time, we would see the universe and all the matter and energy in it shrink into a smaller and smaller space, until eventually the universe shrank down to an infinitesimal point so small that we couldn't see it, and beyond which time as we know it would not exist—a point that physicists call a singularity.

Now imagine that we stop rewinding and press play again. At first we don't see anything on our screen, just blackness. Then, suddenly, out of that infinitesimal point we see a brilliant flash of light as matter and energy pour out, expanding and expanding until we see the universe we recognize today. This brilliant flash, this spark, this flowering of matter and energy at the birth of our universe, has come to be known as the Big Bang.

Georges Lemaître, a Belgian Roman Catholic priest and physicist, was the first to describe a theory similar to what is known today as the Big Bang model. This model points to the universe having a definite beginning. "The evolution of the world can be compared to a display of fireworks that has just ended: some few red wisps, ashes and smoke," Lemaître wrote. "Standing on a cooled cinder, we see the slow fading of the suns, and we try to recall the vanished brilliance of the origin of the worlds."

But this notion that the universe had a beginning did not sit well with many at that time. Some were even disgusted by it. Renowned English astronomer and physicist Arthur Eddington said that "philosophically, the notion of a beginning to the present order of Nature is repugnant." Notice that Eddington didn't claim the science was bad or that these new scientific discoveries were repugnant. Rather, he was bothered by the *philosophical* implications of the discovery.

In fact, it was an opponent of the theory, Sir Fred Hoyle, who coined the name for Lemaître's theory, referring to it mockingly as "this big bang idea" during a radio broadcast. Over the years, many theories were proposed to get rid of this "definite beginning" of the universe, and a war of theories was waged until, finally, in the 1960s, a dramatic discovery left only one winner standing.

Cosmic Static

A PHOTON is a packet of electromagnetic energy. When you turn on a light switch, the light bulb emits an outward flow of light packets, or photons, in the frequency range humans can perceive with their eyes. This allows you to see everything in the room. Other photons exist in frequency ranges invisible to the human eye. But just because you can't see something doesn't mean it isn't there.

On May 20, 1964, at Bell Labs in Holmdel, New Jersey, Robert Wilson and Arno Penzias detected a message from the universe. It came in the form of static noise and it was coming from all directions. They did all they could to remove all possible noise sources, including cleaning off pigeon poop found in the radio antenna. But the noise remained.

Figure 2. The 15-meter Holmdel horn antenna at Bell Labs,
used by radio astronomers Robert Wilson and Arno Penzias to
discover the Cosmic Microwave Background Radiation.

Penzias later remarked, "It wasn't until we exhausted every possible explanation for the sound's origin that we realized we had stumbled upon something big."

What they discovered were photons permeating the entire universe, photons that had only one possible origin, the Big Bang. These photons are known as the Cosmic Microwave Background Radiation.

Some of you may remember an old TV your grandparents had that displayed static noise on the screen when it wasn't tuned properly to a channel. Perhaps as much as 1% of that noise is the result of photons from the Big Bang. Every cubic centimeter of space has at least 300 of these photons moving through it.

Wilson and Penzias's finding corroborated the Big Bang theory and earned them the Nobel Prize in physics. The background radiation they discovered is believed to be the very photons present shortly after the Big Bang. In the early stages of the universe, its size was only about a

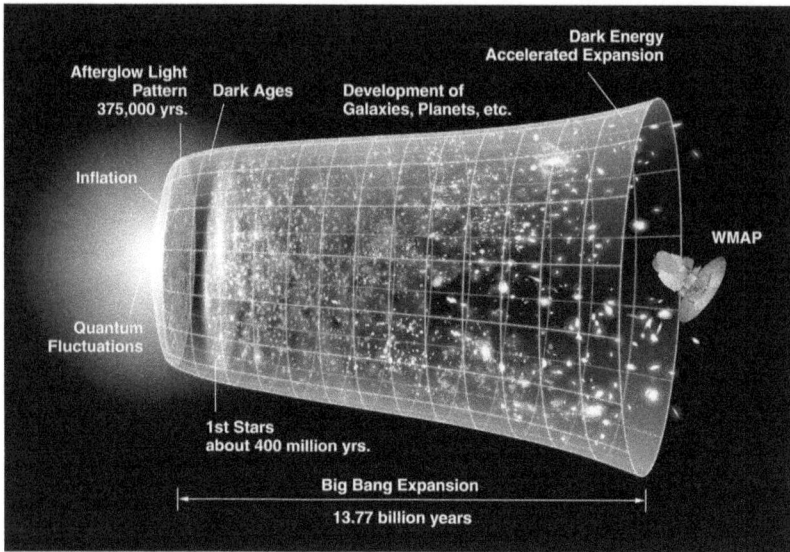

Figure 3. Graphical representation of the development and expansion of the universe during its multi-billion-year history. Note the Cosmic Microwave Background Radiation image (Afterglow Light Pattern) represented near the far left of the graphic, shortly after the beginning of the universe.

hundred-millionth the size it is today, and its temperature was extreme, approaching 300 million degrees. Hydrogen couldn't form because electrons and protons quickly broke apart due to the high speed of the collisions. As a result, traveling photons would scatter off the electrons as light scatters in the midst of fog, without dispersing any farther. It took 380,000 years after the Big Bang for the universe to cool enough for hydrogen to form, thus freeing photons to travel in straight lines.

The Cosmic Microwave Background Radiation is a snapshot of the density of photons at that time—a picture of their distribution through the universe after the electrons combined with protons to form hydrogen and the mist was cleared. Sometimes referred to as the Cosmic Rosetta Stone, it has characteristics that give us many clues about the origin and earliest stages of the universe. These photons could only exist in the evenly distributed state we find them in if the universe were once smaller in volume, denser, and thus hotter.

The Big Bang theory won the war of theories and soon became a household name. But there was more to come. Still more evidence supporting the Big Bang theory arrived as our telescopes continued to advance.

A Video Link to the Past

DID YOU know that you have the power to look into the past? No, I don't mean looking at your memories, or at photos of the past. I mean you actually have the power to look at something as it appeared in the past, not in the present.

Look out your window. If you see the sun (please don't look directly at it; I don't want you to harm your eyesight) you aren't seeing the sun as it presently appears, but rather how it looked about eight minutes ago. Or if the moon is up, take a look at it. What you see is not how it presently is but how it was 1.3 seconds ago.

Or next time it's a clear night, see if you can find the brightest star as seen from Earth, Sirius. You are viewing the star not as it is now, but exactly as it was 8.6 years ago.

How is this possible? Although light travels extraordinarily fast, it isn't infinitely fast. It moves at a little over 186,000 miles per second through empty space. That means it takes photons of light from our host star, the sun, just over eight minutes to reach the earth. Sirius is much farther away—about a half a million times farther away—many trillions of miles from Earth, such that its light, as fast as light is, takes fully 8.6 years to reach us.

A light-year is the distance light can travel in one year. This means that if a star precisely 1,000 light-years away exploded (in the common astronomical idiom, "went supernova") precisely 1,000 years ago next week, we'd first witness that spectacular event in the night sky next week.

On a clear night in the countryside, at the right time of year, you can see the Andromeda Galaxy with the naked eye. What looks like a somewhat hazy star is really the central bulge of an entire galaxy, about

2.5 million light years away. What you're seeing when you look at it is essentially a video feed of that galaxy from 2.5 million years ago.

With our massive telescopes we can look even farther into the universe, much farther away than the Andromeda Galaxy, which is the nearest large spiral galaxy to our own. We can view bright celestial objects many billions of light-years away. And the deeper we look into space, the further we look into the past. What we see at the extreme are portions of the universe in their early stages, baby galaxies as they were evolving into their present-day forms. These results also fit well with the Big Bang model.

All of this spells trouble for anyone wanting to keep the idea of a cosmic creator out of the picture. What is known as the Kalam cosmological argument provides part of the reason why.

Kalam

THE KALAM cosmological argument, popularized by philosopher William Lane Craig, has two premises and a conclusion:

> Whatever begins to exist has a cause.
>
> The universe began to exist.
>
> Therefore, the universe has a cause.

Is the first premise true? "Whatever begins to exist has a cause." Ask yourself, have you ever seen anything come into existence without a cause? No. And in this case, science and common sense have long agreed—whatever begins to exist does indeed have a cause.

How about the second premise? "The universe began to exist."

We saw above how scientific discoveries have put to rest the idea that the universe is infinitely old. With Einstein's theory of general relativity, the insights and observations provided by Lemaître and Hubble, and the discovery of the Cosmic Microwave Background Radiation, along with other evidence, we can reasonably conclude that the universe began to exist. As leading cosmologist Alexander Vilenkin has stated, "With the proof now in place, cosmologists can no longer hide behind the possibility of a past-eternal universe. There is no escape: they have to face the problem of a cosmic beginning." And Vilenkin, it's worth noting, is not religious and has looked for ways to avoid the theistic implications of the Big Bang. To his credit, he has refused to deny the powerful evidence for a cosmic beginning.

Since the two premises of the Kalam cosmological argument are correct, we can safely draw the conclusion: "Therefore, the universe has a cause."

So, what can we infer about the cause of the universe, about what brought it into being? Remember that when the universe began, the space and time of our universe came into being as well, since those too are part of the fabric of our universe, as demonstrated by Einstein's general theory of relativity. So whatever caused the universe to begin must transcend our universe, existing beyond the limits of the time, matter, and space of our universe. And it must be powerful enough, of course, to cause the entire vast universe to begin.

The cause appears to require another important capacity: the ability to fine-tune.

May the Forces Be with You

HAVE YOU ever gone hiking in a forest? If so, more than likely you found yourself surrounded by a still calmness only trees can bring. Birds were possibly singing and squirrels may have been scurrying about. The forest may have felt so peaceful that it appeared as though nature performed this work easily. But the vastness of life surrounding you was being sustained by physical laws so precisely tuned that if they varied from their current strengths even a little, neither you nor the life surrounding you could exist.

With the birth of our universe some 13.8 billion years ago came not only matter and energy but a remarkable set of laws that govern how everything behaves. So delicately balanced are these laws, constants, and initial conditions that small changes would be catastrophic to life. That is, they seem to be fine-tuned to allow for the existence of life. This realization has reshaped the way scientists view the universe.

The universe is governed by four fundamental forces: gravity, electromagnetism, the strong nuclear force, and the weak nuclear force. From the grandest to the smallest scale, these forces govern the way physical objects behave, and the strength of each one appears to be finely calibrated to allow for the possibility of life in the universe.

For instance, if the fundamental force of gravity were somewhat stronger, stars would burn hotter, emitting far more sterilizing X-ray and gamma-ray photons, rendering the surface of a planet like Earth uninhabitable. The stars would also burn much faster. "Typical stars would burn out in a matter of years, not tens of billions of years," explain cosmologists Geraint Lewis and Luke Barnes, meaning that our sun would burn itself out too quickly, long before life had time to arise and diversify on one of its planets.

What if gravity were somewhat weaker? Stars might not go supernova at all, Lewis and Barnes note, and even if one did "and spilled its material into interstellar space, it would be mainly heavy elements such as silicon and iron, rather than life-supporting oxygen and carbon." No carbon and oxygen, no life.

The electromagnetic force is also fine-tuned. Lewis and Barnes describe a sophisticated calculation by Fred Adams that takes account of the settings of gravity and electromagnetism together. When it's plotted on a two-axis graph, the result is a triangle-shaped zone of stability in a vast sea of joint settings that result in unstable stars. "In this figure," the *Fortunate Universe* authors explain, "the 'stable stars' region takes up less than one part in 10^{35} of the whole plot." In other words, the stable-star settings are one in a trillion times a trillion times a hundred billion. Fortunate indeed.

Or consider the strong nuclear force. "If we nudge the strength of the strong force upwards by just 0.4 percent, stars produce a wealth of carbon, but the route to oxygen is cut off," Lewis and Barnes note. What if we go in the other direction? "Decreasing the strength of the strong force by a similar 0.4 percent has the opposite effect: all carbon is rapidly transformed into oxygen, providing the universe with plenty of water, but leaving it devoid of carbon."

The focus on carbon is not arbitrary or due to an unwillingness to imagine other live possibilities. No other element appears remotely capable of replacing carbon as the central element in the buzz of information-processing activity essential to all life. Due to the cooperation of electromagnetism and the strong nuclear force, precise energy levels exist which allow helium atoms to fuse together to make beryllium-8 and then carbon. If these energy levels didn't coincide, the particles would fly apart before they had time to form carbon.

This finely tuned carbon resonance was predicted by Fred Hoyle, and after his prediction proved correct, he commented, "A common-sense interpretation of the facts suggests that a super-intellect has monkeyed with physics, as well as with chemistry and biology, and that there are no blind forces worth speaking about in nature. The numbers one calculates from the facts seem to me so overwhelming as to put this conclusion almost beyond question." Hoyle wasn't religious, so much so that he long resisted the evidence of a cosmic beginning. It seems apparent then that the scientific evidence led him to make the above remark, not religious belief.

This is just a quick and partial survey of how the fundamental forces of nature are fine-tuned to allow for life in the universe.

A Golf Ball Universe is a Livable Universe
SOMETHING ELSE that had to be fine-tuned to allow for life: the texture of the universe at the Big Bang. Weird but true.

Imagine you are holding a ping pong ball. Pay close attention to the smoothness of the ball. Now imagine you are holding a golf ball with its uneven, dimpled surface. Now finally, imagine you are holding a jagged, unevenly shaped rock. The universe at its inception had a texture akin to the golf ball, rather than to the ping pong ball or the jagged rock. That is, it wasn't extremely smooth like a ping pong ball, or highly uneven like the jagged rock. Instead it was relatively uniform but with some variation, akin to the surface of a dimpled golf ball.

And a good thing it was. Our golf ball universe is like the baby bear's porridge that Goldilocks ate—just right. If the dips in the texture of the early universe had been deep beyond a certain narrow limit, there would be massive pockets of gravity. Galactic structures would have clumped together, leading to a bedlam of collisions and explosions lethal to life. Alternately, if the texture of the universe were smooth like a ping pong ball, there would have been too little gravitational pull to form galaxies and planets quickly enough, and thus again, no life. And the margin for error is extremely tight. For life to be possible in the universe, the degree of uniformity—not too much, not too little—had to be fine-tuned within a very narrow range.

The Best Drink in the Universe

LET'S BRING the fine-tuning discussion closer to home. Water. Most of your body is water. Most of Earth's surface is covered by water. Water may seem ordinary to you, but it's actually one of the most extraordinary compounds in the universe, and its suite of unusual properties are essential for life. It's a story of extraordinary fine-tuning for life. Noted author and biochemist Michael Denton ably summarizes the many ways it is uniquely suited for life:

> This wonder fluid is fit for life on Earth in an absolutely stunning number of ways. It is fit for the formation of proto-planetary discs, for the formation of the planets, for the formation of the oceans, and for their subsequent preservation.
>
> Water is uniquely fit for the hydrological cycle, the tectonic cycle, and the temperature regulation of the human body. The properties of water play a critical part in the formation of the great oceanic currents, which circulate crucial nutrients throughout the world ocean. Those

currents play a key role in global temperature regulation and moderation and in controlling the CO_2 levels in the atmosphere.

Water is superbly and uniquely fit to dissolve the minerals in the rocks, and her great solvation powers are fit for the circulation of nutrients both in the blood stream and in the oceans. Water's expansion on freezing and its other unique thermal properties preserve large bodies of water in the higher latitudes.

Water is fit for bioenergetics by providing the proton flows that play such a unique and critical role in the generation of cellular energy. Water's transparency to light is fit for photosynthesis. Water is not only the giver of oxygen, but also uniquely fit for its use in human respiration.

Now, someone might say, well fine, but water was inevitable, wasn't it? With all the different elementary particles in the period after the Big Bang combining in so many different ways, weren't we bound to get a compound like water, even if the laws and constants of the universe were somewhat different? No, actually not. Again, if you think about the fundamental forces of nature, if you tweak some of them even a little, you don't get oxygen. You don't even get elements heavier than hydrogen and helium outside of stars without several things being just so. And you can't get anything remotely like water from elements as simple as hydrogen and helium. Additionally, even if a universe with slightly different strengths for its fundamental forces still managed to produce a version of H_2O, how might it be crucially altered if the strong or weak nuclear force, or electromagnetic force, differed slightly? One or more of water's unique, life-essential properties likely would be altered, spoiling the recipe for life.

Fine-Tuning +

ONE COULD easily fill a whole book with examples of how our universe is fine-tuned to allow for life. There are books out there that do just this. What you have above is just the briefest sampling. But even this quick fly-by urges the question: What caused the universe to be fine-tuned in this way?

If you were a space explorer and, after landing on Mars, discovered a tunnel to an underground room with a climate control system with dozens of sliding knobs, and each one was set at just the right position to allow you to live for days at a time in the room—right temperature setting, right mix of gases in the air, right air pressure, etc.—you would reasonably conclude that the knobs had been purposefully fine-tuned for creatures like you. The laws and constants of the universe are fine-tuned to a far greater degree than this to allow for life.

Now, that fine-tuning isn't by itself enough to make life possible, any more than the hypothetical underground room on Mars was enough by itself. Besides the many finely tuned parameters of the universe, life also requires a host of other fine-tuning parameters that we find fulfilled in our particular location in the universe, planet Earth. Such multi-layered fine-tuning suggests planning and purpose. As the late, great Princeton theoretical physicist Freeman Dyson said, "The more I examine the universe and study the details of its architecture, the more evidence I find that the universe in some sense must have known that we were coming."

A Multiverse Mess

There are many other well-established examples of fine-tuning, which are widely accepted, including by researchers who oppose intelligent design. From what we have covered in this mini-book alone—gravity balanced on a razor's edge, the delicate relationship between electromagnetism and the strong nuclear force, the texture of the universe at the Big Bang, and the amazing properties of water—we see that the parameters required for life in the universe are highly improbable. Yet despite the evidence for planning and purpose, some people still argue that we are here by accident, never mind the long odds.

If faced with a situation in which the probability of success was not in your favor, what would you do to ensure success? How would you raise the probability to be in your favor? When faced with the reality of the fine-tuned parameters of our universe, together with a remarkable planet hospitable to life, some scientists have tried to change the equation by proposing that our universe and our hospitable home were essentially inevitable. They do this by proposing the existence of multiple universes, or what is often called the "multiverse."

The multiverse hypothesis proposes that there are many universes outside our own, each one tuned differently. So, for instance, in one the strength of gravity might be ten times greater than it is in ours; in another, gravity might be much weaker. And on and on with the various other parameters. Then, it is argued, if the number of these universes is astronomically high, perhaps even infinite, then the probability of at least one universe existing with all the right parameters for life is likely, and we just happen to be lucky enough to be in it. Indeed, we should expect ourselves to be in such a universe; otherwise we wouldn't be around to notice our good fortune. Somehow, a group of scientists have managed to flip the low-probability, fine-tuned universe into a highly probable universe. Or have they?

The first thing to note is that there is no testable, empirical evidence that any of these other universes actually exist. Even if these multiple universes did exist, how would we ever know? There is no known way of determining if these universes even exist, since we are limited to observing what is in our known universe. Thus, the multiverse is little more than a hypothetical idea. One influential theoretical physicist compared the idea of other universes with different constants to leprechauns and unicorns.

In addition, one must consider what is creating this massive number of universes. There would have to be some sort of universe-generating mechanism. The most common notion of the multiverse is called the Landscape Multiverse, based on a combination of string theory and eternal inflation. The Landscape Multiverse suggests that there are infinitely many island universes arising from some sort of universe-generating landscape, each universe island containing different higher-level physical laws and constants. The only problem is that this "universe-generating landscape" itself requires fine tuning. In an attempt to remove the fine-tuning problem, they have merely moved it from our observable universe to an unobservable, hypothetical multiverse.

In "A Brief History of the Multiverse" physicist Paul Davies observed the following concerning the multiverse hypothesis:

> How seriously can we take this explanation for the friendliness of nature? Not very, I think. For a start, how is the existence of the other universes to be tested? To be sure, all cosmologists accept that there are some regions of the universe that lie beyond the reach of our telescopes, but somewhere on the slippery slope between that and the idea that there are an infinite number of universes, credibility reaches a limit. As one slips down that slope, more and more must be accepted on faith, and less and less is open to scientific verification.
>
> Extreme multiverse explanations are therefore reminiscent of theological discussions. Indeed, invoking an infinity of unseen universes to explain the unusual features of the one we do see is just as *ad hoc* as invoking an unseen Creator. The multiverse theory may be dressed up in scientific language, but in essence it requires the same leap of faith.

Davies is partly right, but partly wrong, I would argue. The multiverse hypothesis is indeed *ad hoc*, a leap of faith untethered from observational evidence. But what of the alternative hypothesis—a designer and maker of the matter, energy, space, time, and finely tuned laws of our universe? Our uniform experience shows us only one type of cause with the demonstrated power to purposefully tailor and arrange parts in a sophisticated fashion—intelligent agency. We see minds doing this all the time, producing software codes and cars, planes, drones, and satellites, coffee makers and calculators; on and on the list could go. And we never find other types of causes doing so. The fine-tuning of the universe at the Big Bang is such an example of parts tailored and arranged to fulfill a purpose. Inferring a creative intelligence for the fine-tuning of the universe, then, is not *ad hoc*. It is an inference to the best explanation, a tried and true mode of reasoning in the historical sciences.

Review: Your Turn

1. Why did Einstein change his equation?

2. Why did he eventually call it the biggest blunder of his life?

3. Who was Edwin Hubble, and what did he discover at the Mt. Wilson Observatory in California?

4. What might happen if gravity were a little stronger than it is? If it were a little weaker?

5. Why do scientists think the universe is fine-tuned to support life?

6. The space, time, matter, and energy of our fine-tuned universe had a beginning. Something caused the universe to come into being. What are some of the characteristics we can infer about the cause for this cosmic creation event?

7. How is the idea of the multiverse used to explain fine-tuning? What problems, if any, are there with the multiverse explanation?

FUEL YOUR CURIOSITY!

Recommended Resources for Further Exploration:

VIDEOS

Stephen Meyer Shatters the Myth of the Multiverse

Michael Denton Explores the Fine Tuning of the Cosmos for Life

The Big Bang: Something from Nothing?

PODCASTS

God Behind the Birth of Science and the Cosmos
Stephen C. Meyer

Fine Tuning in a Nutshell: No Problem
Robert A. Alston

ENDNOTES

1. Cosmologists continue to refine their understanding of the age of the universe, and some recent findings suggest that the current estimate of 13.8 billion years may need to be revised. However, whether or not new studies and measurements end up changing our best estimate of the exact age of the universe, the key point, as discussed later in this chapter, is the fact that the universe is finite in age, and thus had a beginning.

2. Guillermo Gonzalez and Jay W. Richards, The Privileged Planet: How Our Place in the Cosmos is Designed for Discovery (Washington, DC: Regnery Publishing, 2004), 171.

3. Luke Mastin, "The Expanding Universe and Hubble's Law," The Physics of the Universe, accessed November 1, 2019, https://www.physicsoftheuniverse.com/topics_bigbang_expanding.html. Recent research has revived the cosmological constant, but its precise value does not make for a static universe, as Einstein had hoped. Evidence for an expanding universe is stronger than ever. Indeed, the expansion rate of the universe appears even to be accelerating. In any event, the key takeaway from Einstein's experience is that his preconceptions and commitments led him to explain away, rather than follow, the evidence.

4. Luke Mastin, "Georges Lemaître (1894–1966)," The Physics of the Universe, accessed November 1, 2019, https://www.physicsoftheuniverse.com/scientists_lemaitre.html.

5. Georges Lemaître, quoted in George Gamow, The Creation of the Universe [1952] (New York: Dover, 2004), 51.

6. Arthur S. Eddington, "The End of the World: From the Standpoint of Mathematical Physics," Nature 127 (March 21, 1931): 447–53, https://doi.org/10.1038/127447a0.

7. Helge Kragh, "Big Bang: The Etymology of a Name," Astronomy & Geophysics 54, no. 2 (April 1, 2013): 2.28–2.30, https://doi.org/10.1093/astrogeo/att035.

8. Alaina G. Levine, "The Large Horn Antenna and the Discovery of Cosmic Microwave Background Radiation," American Physical Society, 2009, https://www.aps.org/programs/outreach/history/historicsites/penziaswilson.cfm.

9. Nicholos Wethington, "The Switch to Digital Switches off Big Bang TV Signal," Universe Today, February 16, 2009, https://www.universetoday.com/25560/the-switch-to-digital-switches-off-big-bang-tv-signal/.

10. "6 Things You May Not Know about the Afterglow of the Big Bang," Physics.org, accessed November 1, 2019, http://www.physics.org/featuredetail.asp?id=45.

11. A Short History of the Universe, episode 3, "The Photon Epoch," Highbrow Learning Inc., accessed November 1, 2019, https://gohighbrow.com/the-photon-epoch/.

12. See a brief video about the Cosmic Microwave Background Radiation at Piled Higher and Deeper, "Cosmic Inflation Explained," YouTube, video, 3:42, July 22, 2014, https://www.youtube.com/watch?v=_llA2q1rlSg.

13. Martin White, "The Cosmic Rosetta Stone," Martin White (personal web page), University of California Berkeley, Department of Astronomy, November 1997, accessed November 1, 2019, http://w.astro.berkeley.edu/~mwhite/rosetta/.

14. Karl Tate, "Cosmic Microwave Background: Big Bang Relic Explained (Infographic)," Space.com, April 3, 2013, https://www.space.com/20330-cosmic-microwave-background-explained-infographic.html.

15. William Lane Craig, "The Kalam Cosmological Argument," Reasonable Faith, 2015, https://www.reasonablefaith.org/writings/popular-writings/existence-nature-of-god/the-kalam-cosmological-argument/.

16. Alexander Vilenkin, Many Worlds in One: The Search for Other Universes (New York: Hill and Wang, 2006), 176. His analysis rules out various attempts to postulate a beginningless universe, including the idea that the universe has expanded and contracted eternally, in an endless series of Big Bangs and Big Crunches. As he and co-author Audrey Mithani concluded in a 2012 paper, "All the evidence we have says that the universe had a beginning." Audrey Mithani and Alexander Vilenkin, "Did the Universe Have a Beginning?" arXiv.org, April 20, 2012, https://arxiv.org/pdf/1204.4658.pdf. See also a New Scientist article about his conclusions by Lisa Grossman, "Why Physicists Can't Avoid a Creation Event," New Scientist, January 11, 2012, https://www.newscientist.com/article/mg21328474-400-why-physicists-cant-avoid-a-creation-event/.

17. Anil Ananthaswamy, "Is the Universe Fine-Tuned for Life?," NOVA, PBS Online, March 7, 2012, http://www.pbs.org/wgbh/nova/blogs/physics/2012/03/is-the-universe-fine-tuned-for-life/.

18. Carl R. Nave, "Fundamental Forces," HyperPhysics, accessed November 1, 2019, http://hyperphysics.phy-astr.gsu.edu/hbase/Forces/funfor.html.

19. Geraint F. Lewis and Luke A. Barnes, A Fortunate Universe: Life in a Finely Tuned Cosmos (Cambridge, UK: Cambridge University Press, 2016), 108. As they further explain, "If gravity were 1035 instead of 1040 times weaker than the strong force, then the window would close completely. Stable stars would not be possible at all" (109).

20. Lewis and Barnes, A Fortunate Universe, 108.

21. Lewis and Barnes, A Fortunate Universe, 109–10.

22. Lewis and Barnes, A Fortunate Universe, 118. They add a caveat in the next paragraph. Due to some complicating factors in calculating these hypothetical changes, "it may take a slightly higher percentage to totally eradicate carbon or oxygen from the universe." The operative word here, though, is "slightly higher." Even if this turns out to be the case, the degree of strong-force fine tuning for life will remain striking.

23. Fred Hoyle, "The Universe: Past and Present Reflections," Annual Review of Astronomy and Astrophysics (1982): 16.

24. Martin Rees, Just Six Numbers (New York: Basic Books, 2000), 127.

25. Discovery Science, "Water, Ultimate Giver of Life, Points to Intelligent Design," YouTube, video, 8:05, October 17, 2017, https://www.youtube.com/watch?v=e2i0g1sL-X4. For a deeper look, see Michael Denton, The Wonder of Water:

Water's Profound Fitness for Life on Earth and Mankind (Seattle, WA: Discovery Institute Press, 2017).

26. For books on the fine-tuning of the universe, see Rees, Just Six Numbers and Lewis and Barnes, A Fortunate Universe.

27. Freeman Dyson, Disturbing the Universe (New York: Basic Books, 1981), 250.

28. See Lewis and Barnes, A Fortunate Universe. For a short online overview, see Jay W. Richards, "List of Fine-Tuning Parameters," Discovery Institute Center for Science and Culture, January 14, 2015, https://www.discovery.org/m/securepdfs/2018/12/List-of-Fine-Tuning-Parameters-Jay-Richards.pdf.

29. Simon Friederich, "A New Fine-Tuning Argument for the Multiverse," Foundations of Physics 49 (2019): 1012, https://doi.org/10.1007/s10701-019-00246-2.

30. See, for example, physicist Frank Tipler's discussion in the online video segment, Science Uprising: Fine Tuning (6:27), https://scienceuprising.com/fine-tuning/.

31. Friederich, "A New Fine-Tuning Argument for the Multiverse," 1012.

32. Friederich, "Fine-Tuning," Stanford Encyclopedia of Philosophy, https://plato.stanford.edu/entries/fine-tuning/.

33. Robin Collins, "The Teleological Argument: An Exploration of the Fine-Tuning of the Universe," The Blackwell Companion to Natural Theology [2009] (Chichester, UK: John Wiley & Sons, 2012), 264.

34. Paul Davies, "A Brief History of the Multiverse," The New York Times, April 12, 2003, https://www.nytimes.com/2003/04/12/opinion/a-brief-history-of-the-multiverse.html.

IMAGE CREDITS

Figure 1. Albert Einstein. Photograph by Ferdinand Shmutzer, 1921. Modified by Quibik, 2012, Wikimedia Commons. Public domain.

Figure 2. Holmdel horn antenna at Bell Labs. Photograph by NASA, 1962. Public domain.

Figure 3. Big Bang Expansion. "Timeline of the Universe." Image by NASA/WMAP Science Team. Public domain.

WHAT IS THE DISCOVERY SOCIETY?

The Discovery Society is a group of individuals who come together to support the work—and disseminate the message—of Discovery Institute's Center for Science and Culture. New members receive materials that help educate themselves and spread the word about our work to those in their circle of influence. Depending upon their giving level, members receive one to three Discovery Institute Press newly released books per year, along with invitations to regional donor events and discounted rates on our annual Insiders Briefing events.

If you appreciate this booklet and aren't already a member, we hope you will consider joining our network of supporters today!

Your donation to Discovery Institute's Center for Science and Culture will allow us to expand our cutting-edge scientific research and scholarship; train young people through our education and outreach; and reach the masses through media and communications.

discovery.org/id/donate

MORE INFORMATION ON THE DISCOVERY SOCIETY CAN BE FOUND AT
discovery.org/id/donate/#member-levels.

EVOLUTION AND INTELLIGENT DESIGN IN A NUTSHELL

Are life and the universe a mindless accident—the blind outworking of laws governing cosmic, chemical, and biological evolution? That's the official story many of us were taught somewhere along the way. But what does the science actually say? Drawing on recent discoveries in astronomy, cosmology, chemistry, biology, and paleontology, *Evolution and Intelligent Design in a Nutshell* shows how the latest scientific evidence suggests a very different story.

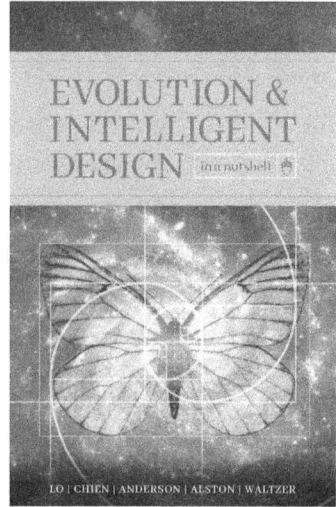

"accessible, informative… powerful … an excellent resource."

J. Warner Wallace

PURCHASE THE FULL BOOK HERE:

DiscoveryInstitutePress.com/EvolutionandID

MORE IN THIS SERIES:

THE ORIGIN OF LIFE & THE INFORMATION PROBLEM
in a nutshell
ERIC H. ANDERSON

FACTORIES THAT BUILD FACTORIES
in a nutshell
ERIC H. ANDERSON

EVOLUTION'S IRREDUCIBLE COMPLEXITY PROBLEM
in a nutshell
ROBERT P. WALTZER

BIOLOGY'S BIG BANG: THE CAMBRIAN EXPLOSION
in a nutshell
PAUL K. CHIEN

This series of booklets was created to help Discovery Society members educate themselves about the basic arguments for intelligent design and the critiques of Darwinian evolution. Each booklet presents the content of one chapter of *Evolution and Intelligent Design in a Nutshell*. To help you delve deeper into each subject, we have included a list of recommended resources from our vast library of videos, podcasts, articles, and websites. Members of the Discovery Society can download digital versions of these books through the Discovery Society Community on the DiscoveryU.org platform or purchase physical copies at a discounted rate through Amazon.com.

www.ingramcontent.com/pod-product-compliance
Lightning Source LLC
Chambersburg PA
CBHW022058190326
41520CB00008B/802